THE TRANSITIONING COMBAT ARMS PROFESSIONAL

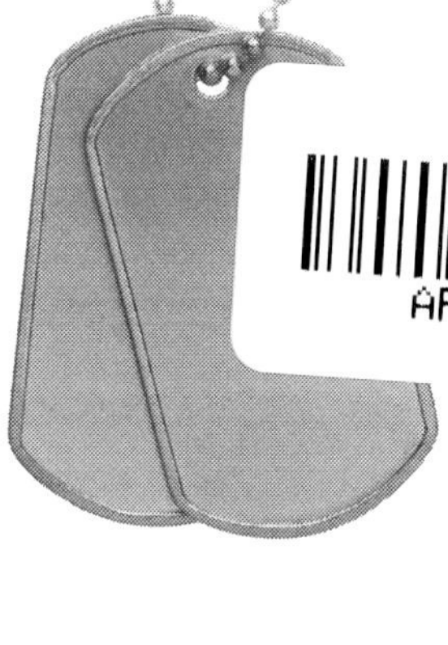

WORKBOOK/COMPANION GUIDE

Jay Hicks, PMP

Lieutenant Colonel, United States Army, Retired

THE TRANSITIONING MILITARY SERIES

GR8TRANSITIONS4U

Published by
GR8TRANSITIONS4U
GR8TRANSITIONS4U (USA) Inc.
PO Box 2
Valrico, Florida 33595

USA

Hicks, Jay
The Transitioning Combat Arms Professional Companion Guide
ISBN 978-1-7343933-3-0
GR8Transitions4U.com

This companion guide is dedicated to You, and to Your successful transition from the United States Military.

Fellow veterans and friends,

Welcome! Thank you so much for reading *The Transitioning Combat Arms Professional!* This companion guide contains assessment questions, star charts, and the Personal Strategic Roadmap to help you consider and apply the concepts contained in *The Transitioning Combat Arms Professional.*

We hope you take time to enjoy your transition and that the book and companion guide prove invaluable

All the best,

Jay Hicks

"The opportunity to secure ourselves against defeat lies in our own hands…"

~ Sun Tzu 孫子

Purpose

A SUCCESSFUL TRANSITION IS NO ACCIDENT. It's the result of honest reflection and focused intentional alignment. While establishing self-improvement goals, you must gauge your progress over time. By building regular checkpoints for self-evolution into your Personal Strategic Roadmap, you are far more likely to stay on a track and find a meaningful post military career.

Set aside some time each day to consider your self-assessments and your way ahead. You will then be able to plot a course of action to take you to the finish line. As you accomplish your goals and objectives, your confidence will increase for resume writing and interviewing, as your "kit bag" will have more tools and capabilities.

The Companion Guide for *The Transitioning Combat Arms Professional* consists of the following elements:

- Examples of assessments, gems, star charts, and a roadmap.
- Assessment translators, enabling you to organize and translate your assessment responses to the gems.
- Large print illustrations of the assessments, gems, star chart and roadmaps. You should print these out and keep them nearby. As you read *The Transitioning Combat Arms Professional*, take notes, conduct your assessments, and plan your transition by creating your Personal Strategic Roadmap.

Refer to *The Transitioning Combat Arms Professional* book chapters for detailed understanding and purpose of assessments, star charting, and roadmaps.

Start early. There is much to accomplish during your successful transition. If you have any questions as you go through this process, please don't hesitate to reach out to GR8Transitions4U.com website and Jay@Gr8Transitions4U. email.

Step 1

Complete the five assessments as described in *The Transitioning Combat Arms Professional.* A large print version of each assessment is provided in this guide. Feel free to reuse the assessment print changing your answers as your personal circumstances change. The following chart depicts the assessment and associated chapter from *The Transitioning Combat Arms Professional* for your cross reference.

Assessment Type	Chapter	Topic Areas
Personal Characteristics	2	Leadership, motivation, creativity managing others, personal growth, organizational skills, working with others, visionary
Environmental	2	Family, re-location, financial obligation, retirement objectives, schools, faith
Timing	2	Service goals met, training/certification goals, time remaining, commitment, financial preparedness
Skills	3	Military skills, education, certifications, credentials, jobs
Market Place	4	Civil Service, contractor, (DoD), commercial market place, analysis based on income, stress competition predictability, longevity, mental growth, benefits

Assessment 1 - Personal Characteristics Assessment *(Chapter 2)*

1. Personal Characteristics	Strongly Disagree	Disagree	Neither Agree or Disagree	Agree	Strongly Agree
I enjoy working with and being around people.					
I easily apply my skills and referential knowledge when performing tasks.					
I tend to lead tasks when given the opportunity.					
I can perform well in a stressful environment.					
I adapt quickly to changing environments.					
I enoy being part of a team effort.					
I have the ability to learn concepts quickly.					
I enjoy cross training, solving problems and finding solutions.					
I have thoroughly analyzed the specialist/management with regard to my future career path.					
I proactively desire to learn and experience new concepts and trends.					

Assessment 2 - Environmental Factors Assessment *(Chapter 2)*

2. Environmental Factors	Strongly Disagree	Disagree	Neither Agree or Disagree	Agree	Strongly Agree
I have performed a post-military financial analysis, to include the loss of military benefits if applicable.					
I have determined my desired geographic location with regard to such factors as healthy lifestyle, allergies, health care access, hobbies, weather and entertainment.					
I have considered my family's special needs in my transition planning.					
I have analyzed and understand myself with regard to work life balance.					
I have given thought to my future location with regard to military base and/or VA Hospital proximity.					
I have studied transition locations with regard to extended family and transportation hub.					
I have analyzed my transition location with regard to future employment, taxation, real estate cost, and overall cost of living.					
I have considered my spouses occupation and their ability to find work.					
I have taken into account my children's primary, secondary and/or college education requirements.					
My family is supportive of my transition into another career.					

Assessment 3 - Timing Assessment *(Chapter 2)*

3. Timing	Strongly Disagree	Disagree	Neither Agree or Disagree	Agree	Strongly Agree
I am ready to leave the military experience behind.					
I have met my career goals for the military.					
I am enjoying or looking forward to making plans for my military transition.					
My resume has been completed and reviewed by a civilian professional.					
I have established a network of professionals in and out of the service.					
I have saved several months salary for financial sustainment during transition.					
I have successfully attended a local military transition assistance program.					
I have successfully branded myself on LinkedIn, Facebook or with appropriate professional associations.					
I have practiced my interviewing skills.					
I have performed a personal SWOT analysis.					

Assessment 4 - Career Field Skills Assessment *(Chapter 3)*

4. Skills	Strongly Disagree	Disagree	Neither Agree or Disagree	Agree	Strongly Agree
I am looking at potential career fields and starting to focus on one or two.					
I have pursued a certification or degree in my desired career field.					
I have performed work to related to my chosen career field while in the service.					
I am comfortable working with technology and technical people.					
I have studied different occupations within my chosen career field with regard to personal interest.					
I have reviewed different types of certifications and analyzed the best certification for my situation.					
I desire to get an additional certification in an essential area, prior to departing the service.					
I have planned projects or implementations in the military.					
I have performed self preparation and I feel confident in my understanding of commercial functions, services and operations.					
I enjoy the staying abreast of the changing commercial environment and advancements within my chosen industry.					

Assessment 5 - Market Place Assessment *(Chapter 4)*

5. Market Place	Strongly Disagree	Disagree	Neither Agree or Disagree	Agree	Strongly Agree
I am aware of my personal stress tolerance, as it relates to each market place.					
I understand the relationship between job, risks and rewards with regard to the market place.					
I have compared and contrasted the associated benefits of each market place.					
I have studied the workplace nuances associated with each market place.					
I have analyzed how well I would transition into each market place.					
I have looked at the pros and cons of job stability, change and longevity within each market place.					
I have researched career progression within each market place.					
I have discussed market place decisions with my significant other.					
I have analyzed each market place with regard to my short and long term personal goals and objectives.					
I have compared the market places against my personal desires for work life balance.					

Step 2

Now that you have completed the assessments, use the assessment translator to organize and translate your assessment responses to the gems. The example below depicts how you array your answers on each gem, which will be used to create the five pointed star.

SCORING: Using question #, plot answer on axis

Assessment Summary: Skills	
Question #	Assessment Answer
Q1	Agree
Q2	Strongly Agree
Q3	
Q4	
Q5	
Q6	
Q7	
Q8	
Q9	
Q10	

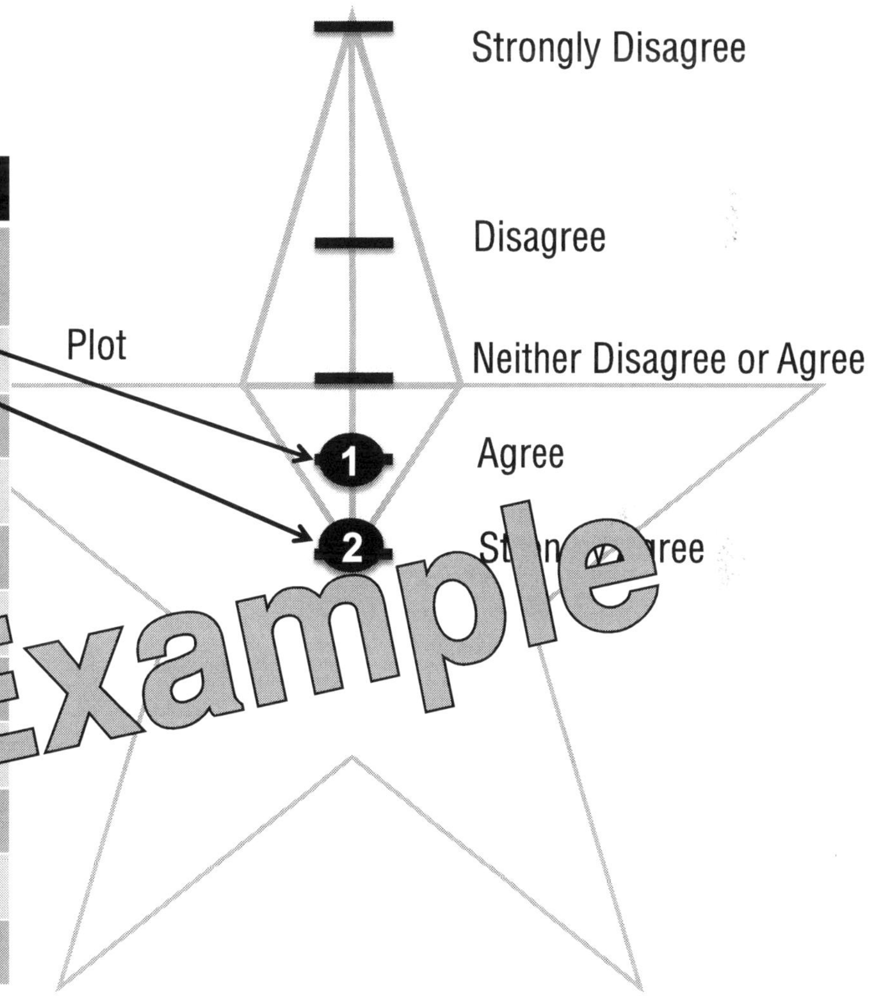

Assessment 1 - Personal Characteristics Assessment *(Chapter 2)*

Assessment Summary: Personal Characteristics	
Question #	Assessment Answer
Q1	
Q2	
Q3	
Q4	
Q5	
Q6	
Q7	
Q8	
Q9	
Q10	

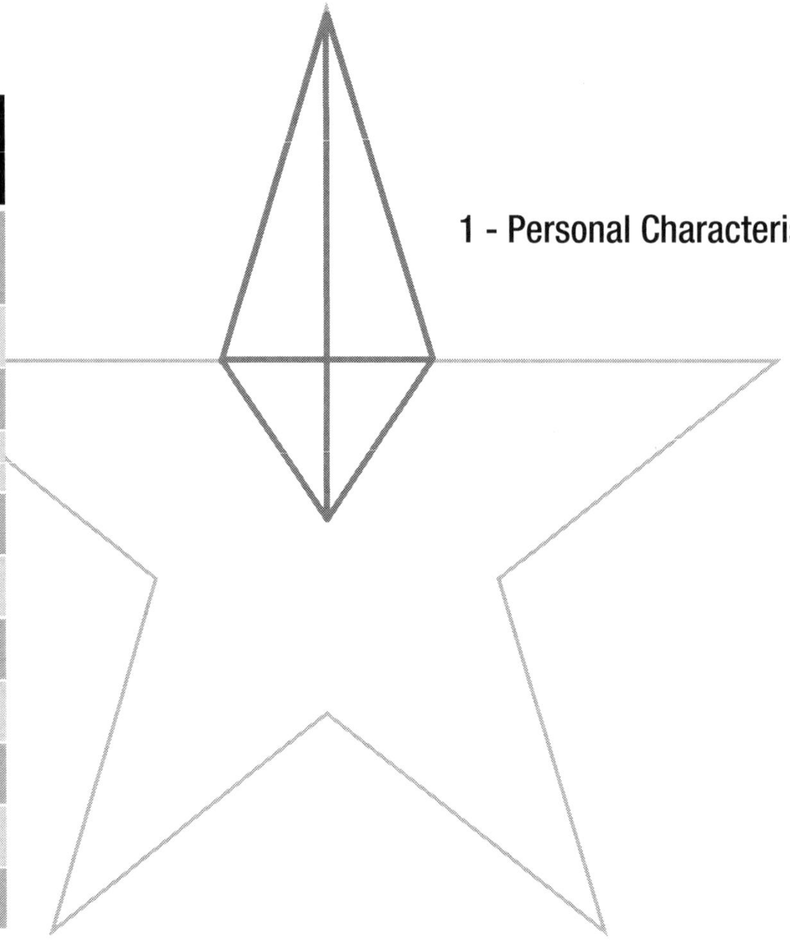

1 - Personal Characteristics

Assessment 2 - Environmental Factors Assessment *(Chapter 2)*

Assessment Summary: Environment	
Question #	Assessment Answer
Q1	
Q2	
Q3	
Q4	
Q5	
Q6	
Q7	
Q8	
Q9	
Q10	

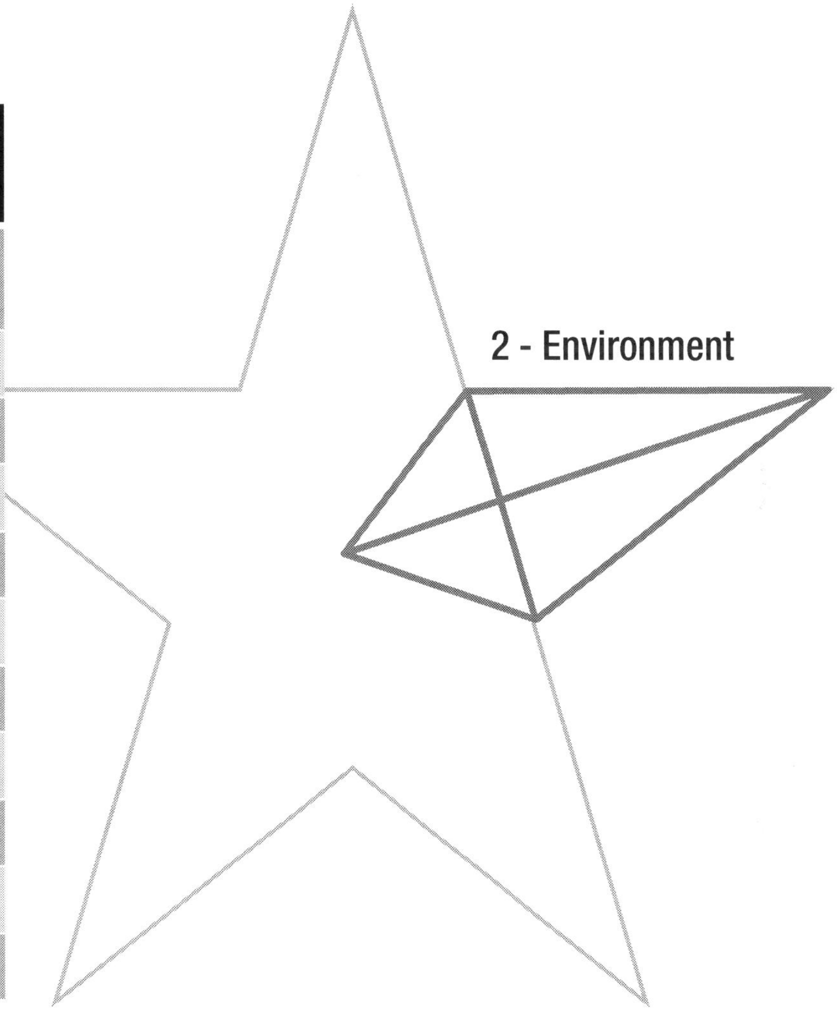

Assessment 3 - Timing Assessment *(Chapter 2)*

Assessment Summary: Timing	
Question #	Assessment Answer
Q1	
Q2	
Q3	
Q4	
Q5	
Q6	
Q7	
Q8	
Q9	
Q10	

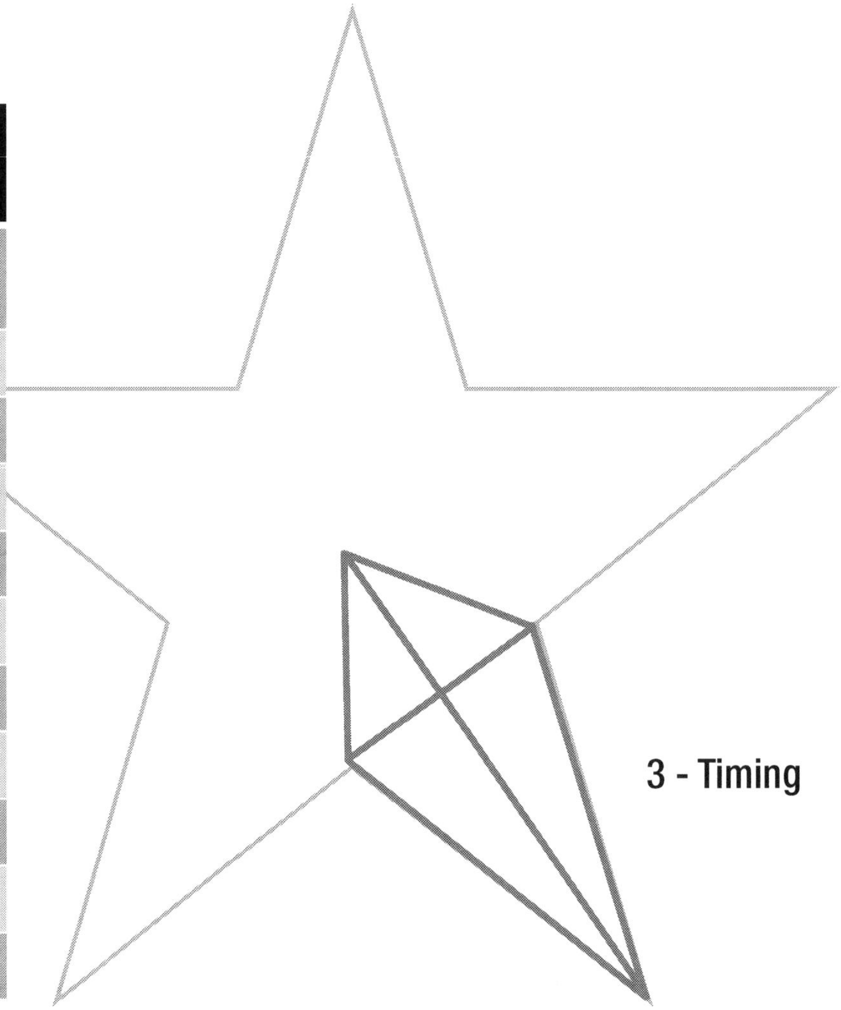

Assessment 4 -Skills Assessment *(Chapter 3)*

Assessment Summary: Skills	
Question #	Assessment Answer
Q1	
Q2	
Q3	
Q4	
Q5	
Q6	
Q7	
Q8	
Q9	
Q10	

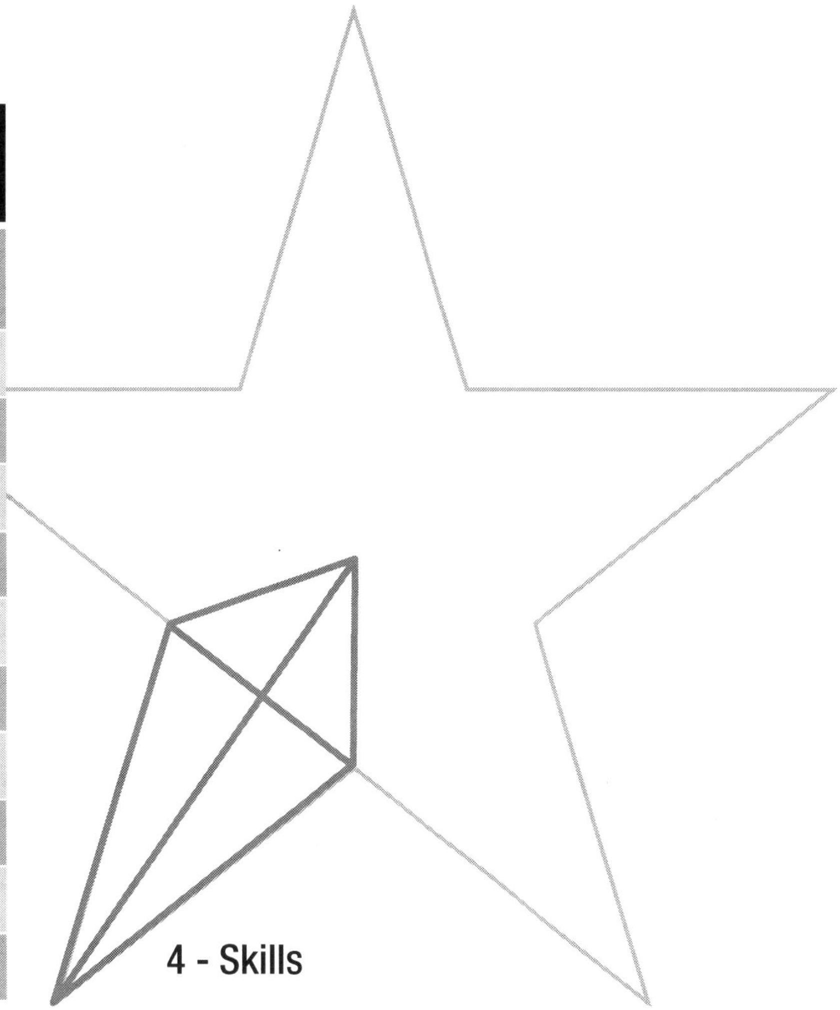

Assessment Summary: Market Place	
Question #	Assessment Answer
Q1	
Q2	
Q3	
Q4	
Q5	
Q6	
Q7	
Q8	
Q9	
Q10	

Step 3

Organize each assessment gem as part of the star, showing your plotted answers in the shape of a five pointed star in the example below.

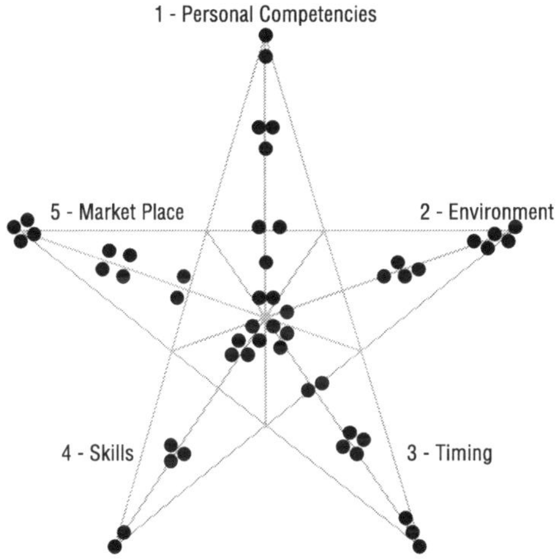

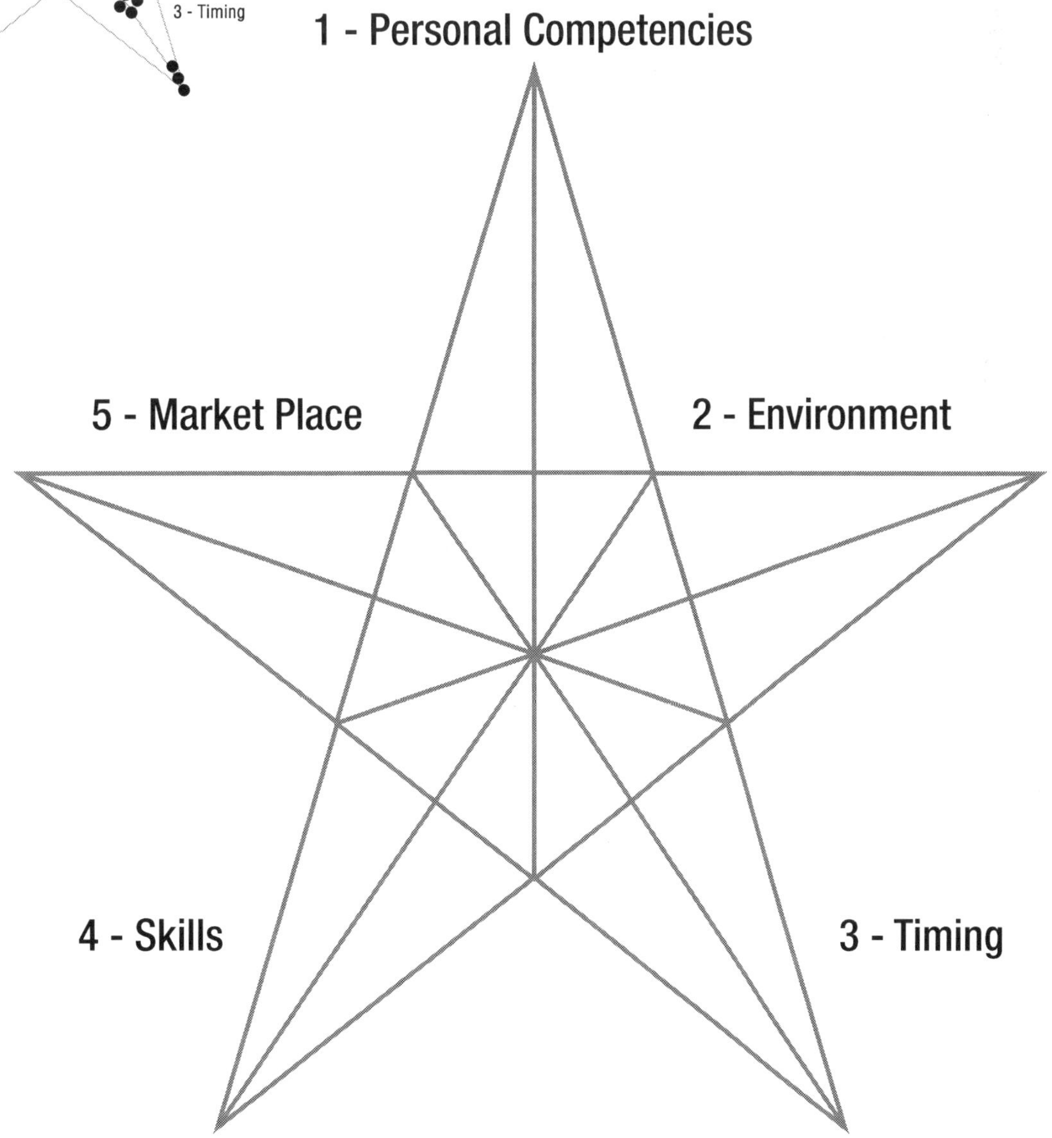

Look for successes and improvement areas falling along the arms of the stars and the pentagon, as described in *The Transitioning Military Project Manager.* Those results within the pentagon shape are qualities that are good for transition. These items should be exploited and can be directly used in either a resume or interview. Results ***outside*** of the pentagon shape are areas to improve upon to increase your success in transition to project management. It is suggested to take 3-4 of these items and transfer them to your Personal Strategic Roadmap (step 4).

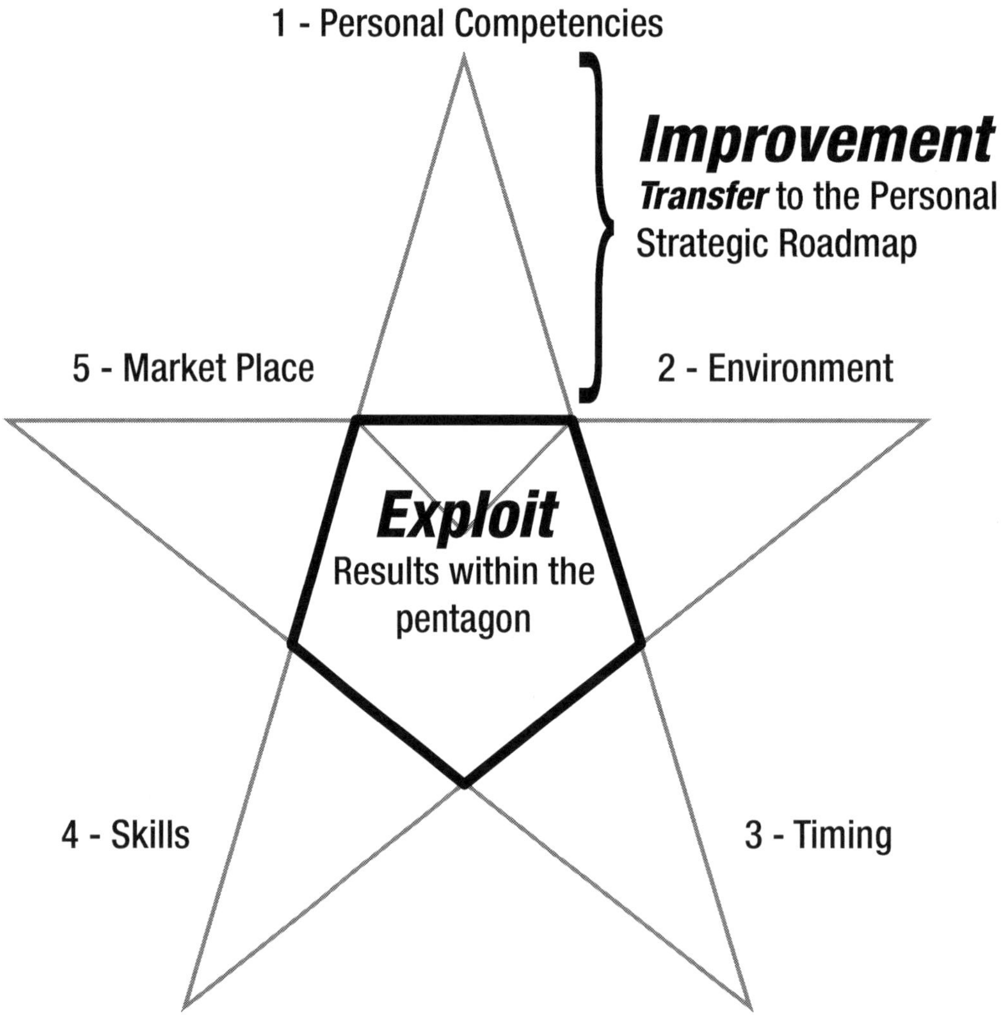

Step 4

The final step is to take the assessment information from the star chart and utilizing the areas that were outside the pentagon. As stated above pick three or four of the most significant improvement areas and establish goals. Reread Chapter 5 from *The Transitioning Military Project Manager* to assist in developing your roadmap.

There are four main sections to the roadmap as follows:

Personal Strategic Roadmap Sections	
1.	Improvements
2.	Vision
3.	Goals and Success
4.	Commit and Attest

Follow the instructions below to complete each section of the roadmap.

Section 1 - Capture Improvements:

Transfer your areas for improvement to the top portion of the roadmap in the improvements section.

Personal Strategic Road Map v1.0

IMPROVEMENTS			
	Vision: Obtain A Project Management career th		
	Mission: Develop a course of action to create a satisfyin		
	ASSESSMENT RESULTS ➡		Areas for Change/Improvemen **Change / Improvement #1**
	1. Environmental/Spousal		
	2. Characteristics		
	3. Timing		
	4. Skills		Need to get my PMP
	5. Market Place	❑ Commercial ❑ Contract ☒ Civil Service	

Section 2 - State your personal Vision:

Based upon your reading and the improvements captured from Section 1, reflect upon how you desire to work on these improvements and state your personal vision. Again, refer to the *The Transitioning Military Cybersecurity Professional* for more detail.

VISION	
	VISION: Based upon your reading and the five assessment areas above, restate your job objectives to include some or all of the following: Job place/location (CONUS/OCONUS, state/city, timeframe, marketplace focus, salary range, possible positions, risk level you are willing to take, industries, and any other considerations.
	YOUR VISION: Make myself marketable for a project management career in the civil service market place in the northeastern region of the U.S., with a major airport nearby. By mid-2015, I want my resume to reflect my credentials and all my project experience.

Section 3 - Create Goals and identify Success:

Now you will create goals from your areas of improvement. If there are more than four improvements, it is recommended that you focus on three or four most important to you. Take into consideration the level of effort and time associated with the success of achieving the goal. Take improvements listed in Step 1, and transfer them to the GOAL STATEMENT section. Check in the ASSESSMENT TIE from what assessment gem you have captured the goal. Post the steps you need to take to achieve the goal in the section PATH TO ACHIEVING GOAL. Set a specific timeframe that you want to achieve the goal in the section YEAR/QTR/MO TO ACHIEVE.

GOALS & ACHIEVEMENT

GOALS:
Pick most important improvement areas from above you want to focus on, depending on level of complexity, learning, or duration (i.e., school).
*List: goals, align which Assessment it ties, year/quarter/month expected to achieve, and present your achievement path (how you will get to your goal and possible steps if necessary.

PRIORITY	GOAL STATEMENT	ASSESSMENT TIE	PATH TO ACHIEVING GOAL	YEAR/QTR/MO TO ACHIEVE (personally set)				ACHIEVED?
GOAL #1	*Get PMI PMP® Certification*	__ Environment __ Characteristics __ Timing __ Skills __ Market Place	*I will research military classes for PMP, study, apply for certification, take the exam and pass!*	*Q1-2015* *Class*	*Q2-2015* *Application*	*Q3-2015* *Take Exam*	*Certificate*	YES / NO Reschedule or no longer need Date ______
GOAL #2		__ Environment __ Characteristics __ Timing __ Skills __ Market Place						YES / NO Reschedule or no longer need Date ______
GOAL #3		__ Environment __ Characteristics __ Timing __ Skills __ Market Place						YES / NO Reschedule or no longer need Date ______
GOAL #4		__ Environment __ Characteristics __ Timing __ Skills __ Market Place						YES / NO Reschedule or no longer need Date ______

Section 4 - Commit and Attest:

Once you identify your goals, sign and date the roadmap as shown below. If you have a family member or a mentor you desire to watch and monitor with you, get them to co-sign.

ATTEST

Date of Next Review: ____________

Signature: ____________ Date: ____________

Spouse Signature: ____________ Date: ____________

Gr8MilitaryPM.com

©Copyright 2014

Re-use and Revisit

Set a reminder through your calendar to review your progress on each goal listed. Depending on the urgency of the goal, review your roadmap weekly.

Make a copy of this form and hang it on your refrigerator, or keep it in your wallet.

If for some reason a goal becomes overcome by events, do not see this as a failure. Carefully assess the situation and the circumstances surrounding the reason why the goal is not achievable and take it off the list. Revisit Steps 1 through 4 and transform other improvements as they develop into goals.

Personal Strategic Road Map v1.0

GR8MilitaryPM ®

IMPROVEMENTS

Vision: Obtain A Project Management career that meets or exceeds my expectations

Mission: Develop a course of action to create a satisfying and financially lucrative transitional outcome.

Transition Date:

RoadMap Initiation Date:

ASSESSMENT RESULTS → Areas for Change/Improvement from Star Chart results *outside the pentagon*

	Change / Improvement # 1	Change / Improvement #2	Change / Improvement #3	Change / Improvement #4
1. Environmental/Spousal				
2. Characteristics				
3. Timing				
4. Skills				
5. Market Place ☐ Commercial ☐ Contract ☐ Civil Service				

VISION

VISION:

Based upon your reading and the five assessment areas above, restate your job objectives to include some or all of the following: Job place/location (CONUS/OCONUS, state/city, timeframe, marketplace focus, salary range, possible positions, risk level you are willing to take, industries, and any other considerations

YOUR VISION:

GOALS & ACHIEVEMENT

GOALS:

Pick most important improvement areas from above you want to focus on, depending on level of complexity, learning, or duration (i.e., school).

* List: goals, align which Assessment it ties, year/quarter/month expected to achieve, and present your achievement path (how you will get to your goal and possible steps if necessary

PRIORITY	GOAL STATEMENT	ASSESSMENT TIE	PATH TO ACHIEVING GOAL	YEAR/QTR/MO TO ACHIEVE (personally set)				ACHIEVED ?
GOAL # 1	*enter goal here.....*	__ Environment __ Characteristics __ Timing __ Skills __ Market Place	*enter steps.....*					YES / NO Reschedule or no longer need? Date ________
GOAL # 2		__ Environment __ Characteristics __ Timing __ Skills __ Market Place						YES / NO Reschedule or no longer need? Date ________
GOAL # 3		__ Environment __ Characteristics __ Timing __ Skills __ Market Place						YES / NO Reschedule or no longer need? Date ________
GOAL # 4		__ Environment __ Characteristics __ Timing __ Skills __ Market Place						YES / NO Reschedule or no longer need? Date ________

ATTEST

Date of Next Review: ______________________

Signature: ______________________ Date: ______________________

Spouse Signature: ______________________ Date: ______________________

GrBMilitaryPM.com

The Military Transitioning Series

The Transitioning Military Series helps service members evaluate and understand their potential to transform themselves into a marketable commodity within both public and private sectors. Each career-based book enables the translation of military experience to the commercial world. Read and use each of these books as a reference to guide during your transition. Insight is provided for those seeking the most satisfying job beyond their military career, with real-world success stories. Companion Guides for each book are now also available through GR8Transitions4U.

A unique combination of features offered through this book series include:

- Career Mapping and Translation
- Transitional Preparedness
- Personal Strategic Roadmap
- Commercial Market Exploration
- Individual Assessments

Interested in another career field? Check out our other books on career field transition for the military:

Project Manager

Logistician

Information Technology

Cybersecurity Professional

Jay Hicks and Sandy Lawrence are dedicated to providing insight and guidance for those looking to transition successfully from the Military with the least amount of stress. Both Jay and Sandy speak around the U.S. in support of transition as well as career field insight, and are available for conferences, podcasts, webinars, and training. For more information on upcoming events and new releases, visit: GR8Transitions4U.com.

Notes:

Made in the USA
Monee, IL
04 November 2026

67380049R00018